A Wisley Handbook

Grapes Indoors and Out

HARRY BAKER AND RAY WAITE

Cassell

The Royal Horticultural Society

THE ROYAL HORTICULTURAL SOCIETY

Cassell Educational Limited
Villiers House, 41/47 Strand
London WC2N 5JE
for the Royal Horticultural Society

First published 1979
Second edition 1985, reprinted 1985, 1987, 1988
Third edition 1992

British Library Cataloguing in Publication Data
Baker, Harry
    Grapes indoors and out. – New ed.
    1. Viticulture
    I. title   II. Waite, Ray
    634'.88   SB397.G7

ISBN 0-304-32014-5

Photographs by Harry Baker, Wilf Halliday, Photos
Horticultural, Martyn Rix, Harry Smith Collection,
CRH Photographic
Line drawings by Peter Mennim and Michael
Shoebridge

Phototypesetting by Chapterhouse Ltd, Formby
Printed in Hong Kong by Wing King Tong Co. Ltd

Cover: 'Seyval' is an easy-to-grow hybrid grape.
    Photograph by Harry Baker
p.1: 'Madeleine Angevine' is vigorous and heavy
cropping.
Back cover: the famous 'Pinot Noir', a high quality
wine grape.
    Photographs by Harry Smith Collection

Contents

Vines under Glass

During the Victorian heyday of the private estate and even to some extent into the early part of the twentieth century, vines were grown on their own in special greenhouses, called vineries. In some large privately owned gardens there were even separate greenhouses for early, mid-season and late fruiting cultivars but it was more usual to divide the greenhouse into sections each with a different growing temperature.

After the war the days of the specialized vinery were ended and relatively few people grew vines under glass. Recently, however, interest has re-awakened and now more amateur gardeners are finding that they can grow a worthwhile crop of grapes in the roof space above other plants, even in a small greenhouse. Growing vines in pots can be even more effective, a small greenhouse 8 feet × 6 feet (2.4 × 1.8 m) giving good results.

Most dessert grapes need the protection of glass to ripen properly and achieve their full potential. Certain cultivars, notably the Muscats, need artificial heat in the spring particularly at flowering to aid pollination, and also in the autumn, so that the grapes attain their best flavour. In view of the cost of a greenhouse, the cost of fuel and the amount of detailed work necessary to keep the vines under control, it is better to use the greenhouse for dessert grapes rather than wine grapes which can be grown outside. The limiting factor in a small greenhouse is head room, but the current fashion is for a greater height to the eaves, so that problem may now be less significant.

The ideal structure for grapes is a lean-to greenhouse facing south; the roof rises across the whole width of the house affording as large an area as possible for training the vines, and in addition the rear wall collects a reserve of heat on a sunny day which is released into the structure at night. However, vines can also be successful in free-standing span greenhouses.

Vines can be grown with or without heat. The latter provides a later crop, but limits the number of cultivars that can be successfully ripened. In a heated greenhouse the vine can be started into growth in mid-February, and ripe fruit expected in July and

---

Opposite: 'Mrs Pearson' (see p. 27)

August, depending on the cultivar grown. In a cold greenhouse, growth will start naturally in March, with harvesting in late August and into September.

A vine grown without heat can be planted out of doors and the main stem (called the rod) trained through a hole made low down in the greenhouse wall, or it can be planted inside with the roots allowed to grow out in the soil outside.

Vines grown in heated greenhouses are best planted in a specially constructed border. With roots inside, the grower has greater control and success or failure depends on his or her knowledge and judgement. The method is of particular advantage in the colder parts of the country where an early start is required or if there is any risk of the roots growing into areas which might become waterlogged in winter. In the past it was always said that for grapes started very early in the year the inside border ensured that the root action was on the move at the same time as the warmed top growth. However, this is a counsel of perfection and for most needs today outside planting will be satisfactory, making for easier initial preparation and less frequent attention to watering and feeding.

PREPARATION FOR PLANTING

Whether planting is to be done outside or inside the greenhouse, soil preparation means double digging, incorporating well rotted manure or compost, together with a dressing of general fertilizer such as Growmore. Bonfire ash can also be added if available. If there is any suspicion that the soil might become waterlogged construction of a special bed will be necessary. Make the border $2\frac{1}{2}$ to 3 feet deep (75–90 cm) with solidly constructed sides and base, for example brick retaining walls and a concrete floor. The floor should slope from back to front and from one end to the other and a land drainpipe 4 inches (10 cm) in diameter be laid along the lower edge with an outlet at the lowest corner leading to a ditch, surface water drain, or a soak-away prepared by digging a large hole and filling it with stones or brick rubble. Spread similar drainage material to a depth of 6 inches (15 cm) over the whole of the base before filling up the border space with soil. This soil should be capable of holding reserves of water and plant food, at the same time being open and porous. The recipe for John Innes potting compost no. 3 – 7 parts by volume of sterilized loam or fertile garden soil, 3 parts by volume of peat, 2 parts by volume of coarse sand, plus 12 oz (300 g) of John Innes base fertilizer and $2\frac{1}{4}$ oz (55 g) of ground chalk to every bushel (36 litres) – will provide a

As the bed is enlarged in the first years of growth, good drainage material is placed below the compost mixture

satisfactory medium. If wood ash from a bonfire containing charcoal is available, this can also be included in the mixture.

The young plant will not require a very extensive root-run in the first year, and the old practice of making up a narrow border for the first year and adding to the width annually has much to commend it, for it makes less initial work and reduces costs. The method ensures that a medium full of nutrients and in good physical condition is being added year by year during the early life of the vine (see illustration above). Concrete building blocks provide a convenient means of retaining the compost, but ordinary house bricks or planking will serve equally well.

The growing vine will need to be trained inside the greenhouse so preparation must include the fixing of wires for support. It is usual to run the wires along the length of the structure starting at about 3 feet (90 cm) off the ground, and continuing at about 9 inches apart (22 cm) and 12 to 18 inches (30–45 cm) from the glass.

In a small greenhouse it is only feasible to allow 12 to 14 inches (30–35 cm) from the glass, but the young shoots tend to grow vertically and quite rapidly and will soon stub themselves against the glazing and become scorched or distorted if they do not have enough space. It is also essential to keep the air moving between the glass and foliage so the greater the space given to the wires the better. In a larger greenhouse, allow about 18 inches (45 cm).

PLANTING

In the small greenhouse there will probably be room to plant only one vine. This is best done at the opposite end to the door so that the rod can be trained along the whole length of the ridge. Alternatively, the young vine can be planted to one side, at the end near the corner, and run along the side wall.

Traditionally vines were planted side by side and this method lends itself perfectly for the large greenhouse especially of the lean-to design. Plant the rods at 4 feet apart (1.2 m) as the spread of lateral growth can be expected to reach 2 feet or more (60 cm) either side.

Young vines are usually sold in containers and therefore can be planted at any time of the year, but November and December, when the plants are dormant, are the best, as at that time the main stem can be cut back by two-thirds without fear of excessive sap 'bleeding'. This severe pruning ensures vigorous growth in the summer and also stimulates the lower buds to form side shoots (laterals), so encouraging lower spur systems to be formed. If plants are not obtained until later in the winter, it is unwise to prune them, otherwise the rising sap will bleed for several days from the cut, and considerably weaken the vine. Similarly, vines planted in full growth should not be pruned until after leaf-fall.

It is not always possible to purchase strong vines (stem about $\frac{1}{2}$ inch (1 cm) in diameter) in which case it is better to pot on into a larger pot, reduce the main stem to one-third its original length, and allow the plant to grow for another season before planting in the following winter.

THE FIRST YEARS

In the first year after planting (and after the initial hard pruning) allow the main stem to grow unchecked so that it produces the maximum length of vigorous shoot. If the roots are growing well, the main shoot should reach 10 feet (3 m) or more and produce lateral shoots over much of its length. During the summer these shoots are stopped when they have made five or six leaves, and any sub-laterals (side growths formed on the laterals) produced are cut to one leaf. Both main stem and laterals are loosely tied into position on the wires as they grow, allowing the stems ample room to increase their girth.

At the end of the season and as soon as the leaves have fallen, the main stem is cut hard back again, removing two-thirds of the summer's growth. The lateral shoots are pruned to one plump bud.

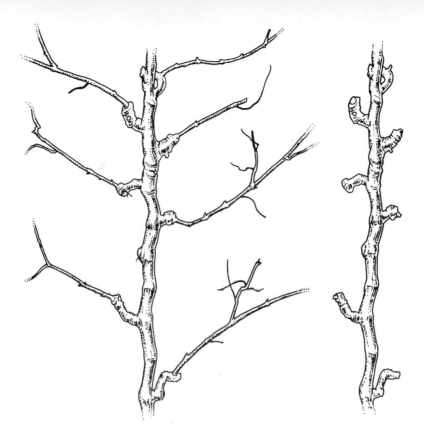

The glasshouse vine, grown on the rod and spur system of pruning – before (*left*) and after (*right*) pruning immediately after the leaf fall

In the following years the same technique is carried out until the main stem, after being pruned by two-thirds of its length, has reached the furthermost wires. Ultimately this form of training will result in what is known as the rod and spur system, a permanent framework consisting of a main stem with the laterals forming spurs at intervals of 9 to 12 inches (22–30 cm) along it. It is from these spurs that the fruiting shoots arise each year. The main stem between the spurs is barren. This is the simplest form of training, but it is also possible to grow two or more rods on one plant or allow the rod to branch and extend until one vine fills an entire greenhouse, as does the famous vine at Hampton Court, London.

WINTER PRUNING OF ESTABLISHED VINES

Vines should be pruned immediately after leaf-fall, which is usually in late November or December. At this time the greenhouse must be quite cold in order to keep the vines dormant. This is done by giving as much ventilation as possible after harvest so that the current year's growth is fully ripened. This hardening of the growth will, with the early winter pruning, avoid the real risk of a harmful level of bleeding when the sap rises in the spring. Bleeding can be prolonged and abundant, and although various 'cures' have been suggested, none are wholly satisfactory. If later pruning is unavoidable, the application of carpenter's knotting to the cuts immediately after pruning may help, but it is better to avoid the trouble rather than to cure it.

With the established rod and spur system, winter pruning consists of cutting back all growth made in the past year (other than that required for extension or replacement) to within one or two buds of the older wood. One bud is enough if it is strong and plump, but often the basal bud is small and unlikely to give fruiting wood. Pruning to two buds will give two growths so that should one become damaged or broken away, there is still one to be tied in to fill the space (see illustration on p. 13).

Indoor vines can also be grown on a replacement pruning system. This requires more skill but will generally result in higher quality fruit. With this system the winter pruning is much the same as that for the Guyot method of growing vines in the open (see p.40). Two new main stems are allowed to grow each year, one of which is retained for fruiting and the other cut back to two buds to produce the two rods for the following year. The rod which has fruited is cut out completely.

WINTER MAINTENANCE

The practice of scraping away the oldest bark from a rod takes time and some growers regard it as unnecessary; but if there is time, it is still well worth doing. Special attention needs to be given to the flaky bark on the spur, but take care near the buds so as not to damage them. Much of the fibrous bark can be rubbed or pulled away but when scraping with a knife, the green tissue must not be exposed, going down just to the smooth brown bark.

By scraping away the old bark, over-wintering pests, such as red spider mite and mealy bug, are either removed or exposed to a pesticide which can be sprayed on, but preferably brushed on ensuring that it is rubbed into crevices and angles of the bark.

If vines are growing in the greenhouse on their own, a 4% solution of tar oil winter wash (30 cc of tar oil in $\frac{1}{4}$ litre of water/1 fluid ounce to $1\frac{1}{4}$ pints) may be applied. The smell, although not harmful, can be persistent; if this is offensive near the house, an alternative is malathion at the strength recommended for spraying – even this, however, has some smell. It is best to complete this work by early January, so that there is no chance of injuring the buds. This cleaning of the rods is more convenient if they are untied and lowered from the wires. Remove all the old ties from the wires as these too may harbour pests.

The next job is to renovate the inside bed. The remains of the old mulch are raked from the bed and the surface soil gently pricked over with a fork. This loose soil is raked or brushed off with a stiff broom or besom. The roots thus exposed are immediately top-dressed with a good loam-based compost such as John Innes no. 3.

No more needs to be done for the next two or three months, until the vine is started into growth. As this time approaches, the soil in the bed is watered thoroughly; a second watering a week later may be needed if the root area has become very dry during the winter. After watering, a mulch of well rotted farmyard manure, spent mushroom compost or similar material should be applied; this will help to regulate the soil moisture during the growing season. This is also the time to extend the vine border in width as necessary.

For grapes planted outside the greenhouse, no special winter treatment is required other than an application of a good general fertilizer, when growth begins.

THE SPRING ROUTINE

If warmth is required for the needs of other plants, heat may be provided in February; the middle of the month is an ideal starting time for vines.

The rods which have been loosely tied to the wires for the winter operations are untied and lowered so that when supported about a third of the way up, the end is bent down, almost touching the ground. In this way the rising sap is prevented from rushing to the top buds and stimulating them into growth at the expense of those lower down. When growth has started on all the spurs, the rods are tied into their summer positions on the wires.

Temperatures at this stage may rise rapidly in sunlight and as a gentle start is required, careful ventilation must be given at about 19°C (66°F). With artificial heat a night temperature of 4 to 7°C (39–45°F) should be the aim. Keep the atmosphere moist by damping down the floor and walls on fine days. Until flowering begins, the rods can be sprayed overhead with clean water, but do this in the morning before the sun becomes too hot, for there can be

11

A flower bunch: the petals have fallen and the young grapes are beginning to form

a risk of scorching the young foliage by the action of the sun's rays through the water droplets.

If it is necessary to reduce the number of growths to give one shoot on a spur, the surplus ones should be removed as soon as possible. In practice, it is safer to pinch back the weaker unwanted shoots to two or three leaves, in case a potential fruiting shoot is accidentally broken when tying in (see illustration opposite).

As the shoots will grow vertically, they must be brought down to the wires; this can be rather tricky as they may easily snap at the junction with the older wood. Growths should be trained down gradually, doing this every other day (see illustration opposite). It is unwise to tie in while the growth is very young and brittle, and better to wait until a little fibrous tissue has formed. The task should be done in the morning, for although later in the day the shoots can be limp and more easily managed, at night they become turgid again and if tied too tightly may break. As the light intensity increases, apply light shading to the glass to prevent leaf scorch. This may need to be done as early as April with a small greenhouse. Shading will also help to reduce excessively high and too rapid rises in temperatures and thus to some exent prevent drying conditions.

FLOWERING AND POLLINATION

By the time the shoots are 18 to 24 inches (45–60 cm) long, the flower trusses will be quite prominent and a start can be made in pinching out the shoot tips. Two leaves are allowed to form beyond the flower bunch before the growing tip is removed. If no bunch has been formed, the shoot is left to make six or eight leaves before pinching. All side shoots that form on the new growth should be pinched at one leaf.

Although some cultivars set their fruit very readily, it is unwise to leave pollination to chance. With freely setting cultivars like 'Black Hamburgh', 'Foster's Seedling' and 'Alicante', all that is needed is to give a brisk shake to the rods. For shy setters like the Muscat types, the easiest effective method is to stroke the bunches gently with cupped hands in order to transfer the pollen from the stamens to the receptive stigmas. If pollen can be transferred from a different cultivar, so much the better. Always try to pollinate at mid-day. At this critical stage, keep the atmosphere on the dry side, and although warmth will help pollination, air must be kept circulating. Even on a cool day, a little ventilation will help. Do not spray overhead on such days.

Removing a surplus shoot from a spur (*left*), and a lateral shoot tied down with a running noose (*right*)

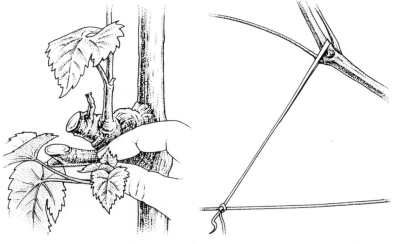

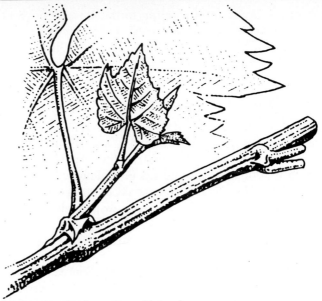

A stopped shoot showing the formation of laterals

THINNING THE CROP

If all the bunches are allowed to mature, the quality of fruit will suffer; in addition the vigour of the shoot growth is lost and fruiting may be upset for several years. The number of bunches left to fruit must therefore be regulated. As a general rule of thumb, a pound of fruit is retained for every foot run of rod (450 g per 30 cm). This lightening of the load should be done as soon as it is possible to assess which bunches show promise of being shapely, well set, of good size and in such a position that they will develop freely.

Next the grapes must be thinned. It is not a difficult task, but takes time. Special grape thinning scissors are obtainable but ordinary small scissors will do, provided they are sharp, particularly at the tip, and are not too cumbersome. A small stick about 6 inches (15 cm) long, forked at one end, is helpful in steadying the bunch. Grapes should never be touched with the fingers, as this damages the bloom on the skin. This damage not only disfigures the bunch, but removes the extra protection given by the water repellent qualities of the bloom.

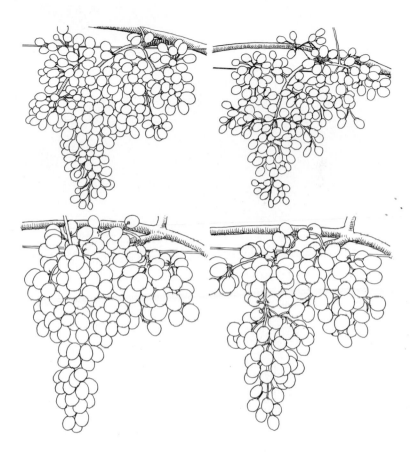

Above: bunches before (*left*) and after (*right*) first thinning
Below: bunches before (*left*) and after (*right*) second thinning

Occasionally it is necessary to shape the bunch before thinning, and this usually involves the removal of a straggling arm at the top of the bunch. Any large bunches may need to have their top branches, referred to as shoulders, supported by looping them up to the wires with raffia. Excess grapes in the middle of the bunch are cut away first, then any undersized berries as these are probably imperfectly set. Next, the outer berries are spaced to allow room for swelling. The shoulders should not be thinned too much as they should stand firm and square when the bunch is harvested.

Thinning can seldom be completed in one operation and it is usually necessary to remove a few more grapes three or four weeks

'Alicante' is a vigorous grape which benefits from good thinning (see p. 25)

after the first thinning. A final thinning may be required after the pips have formed when the grapes undergo a second period of swelling.

WATERING AND FEEDING

A vine in full leaf will make heavy demands on soil moisture in warm weather. The intervals at which water is required will vary according to the climate and soil conditions. If the vine's roots are confined inside the greenhouse, a generous application of water will probably be necessary every seven to ten days, and provided drainage is good, the risk of over-watering will be slight. It is better to water thoroughly so that moisture penetrates several feet into the border soil. If the roots are outside, they may be supplied by natural rainfall for much of the year, but should not be forgotten in prolonged dry spells.

Feeding will be required while growth is active. For good quality fruit the main need is for a high potash feed, and any of the tomato fertilizers available will be ideal. Water on a solution at two- to three-week intervals from about a month after starting into growth until the fruit begins to ripen.

Sometimes the vine is not vigorous – a condition brought about perhaps by overcropping, old age or poor soil conditions, in which case the application of a general feed, i.e. with equal quantities of nitrogen, phosphate and potassium, or one containing higher nitrogen, will be beneficial, but these should only be used from April to May.

RIPENING

Throughout the summer and autumn a healthy vine will grow vigorously. Any new shoots not needed to fill gaps in the foliage canopy or to add to the length of rod for the following year are cut back after one new leaf has formed; this must be done promptly or the growths will rapidly get out of control. Remove tendrils immediately they form.

When the grapes start to change colour, there is a greater risk of splitting. This stage is easily recognised in black or tawny grapes, but is not so apparent with green ones, although a slight change of colour towards white or amber is noticeable.

A damp atmosphere encourages splitting. Damping down of the path and other surfaces should be done by the middle of the day so that any excess moisture has had a chance to dissipate before night. If there are other plants in the house, they should also be watered well before evening. Leaving the lee side top ventilator open for an inch or two (about 3 cm) will help to dry the atmosphere and will also control to some extent any early build up of heat in the morning. The air temperature should never be allowed to rise rapidly, because the temperature of the grapes rises slowly, and moisture from the air will condense on the fruits which may then split. If the soil is allowed to get very dry, the fruit skins will harden and when water is finally given, they are likely to split. A thin mulch of straw applied over the soil when the fruit has formed will help to

'Chasselas Rose Royale' (see p. 49)

keep the soil uniformly moist and the air in the greenhouse on the dry side. From the early stages of ripening, look over the bunches two or three times a week for diseased grapes, removing those with a split skin; continue this until the end of the season. Cut out the bad grapes using scissors and the forked stick, working with care so as to keep the bloom intact.

Grapes are not completely ripe when fully coloured and require 'finishing'. This means allowing them to hang for a period so that sugars are formed. The finishing period varies with the cultivar and time of year. 'Black Hamburgh' and 'Foster's Seedling', ripening in the summer, need only two or three weeks to mature fully; 'Muscat of Alexandria', 'Mrs Pince' and 'Syrian', which ripen later, need to hang for eight to ten weeks before they are at their best.

'Huxelrebe' (see p. 49)

When harvesting the grapes, handling will be made easy if the branch carrying the fruit is cut with a 'handle', i.e. leaving about 2 inches (5 cm) on either side of its joint with the main stem. This handle allows the bunch to be inspected, carried, mounted on a showboard or placed in a dish without disfiguring the fruit.

Ripe grapes may be kept in good condition on the vine for several weeks by using artificial heat and ventilation to maintain a cool dry atmosphere (about 7°C/45°F).

If a cool, even temperature can be provided in a room, it is worth cutting the late grapes with a stem in December. The stem can be placed in a jar of water and in this way fruit will keep for a considerable time. 'Lady Downe's Seedling' is particularly good for this purpose.

19

Vines in Pots

The cultivation of vines in pots has several advantages. The containers can be readily moved within the greenhouse and moved outside after fruiting. Such outdoor treatment will ripen the shoots and give the cold environment necessary for satisfactory dormancy. Clay pots will need to be plunged to reduce the chance of frost damage to the containers, but whatever the receptacle, plunging gives stability in windy weather.

Much of the work described for vines under glass applies equally to the management of pot-grown grapes except that they are grown as standards, with a long stem producing a 'head' of shoots (see illustration overleaf). The method deserves to be more widely used because worthwhile crops can be obtained, especially in the smaller greenhouse. Where space is limited, more than one cultivar can be grown and with care, the more usual crops can be cultivated at the same time. The vines are taken out of the greenhouse in the autumn, leaving room to house chrysanthemums, for example. Heating will give a long season of growth and therefore earlier ripening, but an unheated greenhouse can also give good results.

Early fruiting cultivars will be found to be the most satisfactory with 'Black Hamburgh' still possibly the most suitable; but 'Buckland Sweetwater', 'Foster's Seedling' and the Frontignan types all give good results.

Well established rods should be obtained and are best potted just as the roots become active in the spring. Weak rods will need to be grown on so as to build up the main stem and it will be necessary to cut the rod back to one bud so as to induce vigorous growth during the summer. Any pruning should be done during November and December. Rods with a diameter of $\frac{1}{2}$ inch (1 cm) or more can be pruned to a bud (eye) at 3 to 5 feet (0.9–1.4 m) above the pot. The length of stem chosen will depend on the height of the greenhouse.

Vines may need to be potted into a larger size pot as the root system grows, and this is best done as the roots start into active growth in the spring. A vine in a 5 inch (13 cm) pot should be potted on into a 7 inch (18 cm) pot but plants with larger root balls can be moved into a 9 inch (22 cm) pot. A loam-based compost, e.g. John

'Black Hamburgh' has a good flavour and does well in containers (see above, and p. 25)

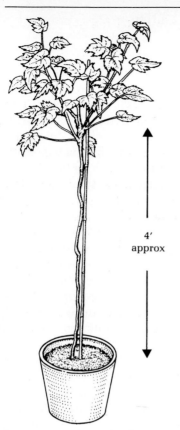

4′
approx

'Buckland Sweetwater', showing the new
shoots arising from pruned laterals

A pot-grown vine trained as a standard

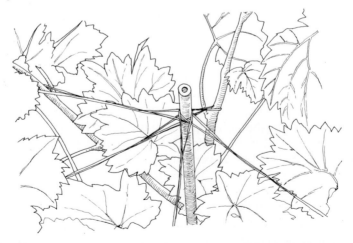

Branches are secured to the central cane with raffia or garden twine

Innes no. 3, is best, because it will ensure that the plant will not become top heavy.

Once started into growth, the shoots soon lengthen and the top four or five shoots should be left to develop a 'head', being pinched when they have made five or six leaves. Any growths on the stem should be removed completely. Flower clusters are unlikely to appear but if they do, remove them, as they should not be allowed to develop at this early stage.

For the second year the ripened shoots should be pruned in the dormant season to one bud. This is the start of building up the spur systems. These will form the permanent head of a pot-grown vine. As active growth begins in the spring, scrape off the existing top layer of soil then add a top dressing with John Innes potting compost no. 3 to the top of the pot, being careful to leave enough depth for thorough watering. To encourage strong growth and to ensure that it is well ripened stand the vine out of doors from mid-summer onwards.

During the third year prune each shoot hard to one good bud in early winter. A move into a larger container will almost certainly be required, but this is likely to be the last move as with top dressing and feeding, a vine can remain in the same pot for many years.

The vine can produce its first fruit in the third year. Pinch the growths at two leaves beyond the bunch, and any laterals and sub-laterals at one leaf.

The growths will need support and this can be done by tying them to a central cane in a maypole fashion. A general liquid fertilizer is best applied every seven days up to fruit set, after which a high potash feed (e.g. a tomato fertilizer) should be used.

In warm situations, it may be possible to finish ripening the fruit outside, but it will need to be protected from damage from birds and wasps, by covering with netting, or enclosing the bunches in nylon stockings.

Although a vine grown as a standard is a convenient method of training, a simple rod and spur system can also be adopted. In this case the lower shoots may be allowed to develop and should then be pruned back in subsequent years, as described for permanently planted vines (see p.34).

Cultivars for Growing Under Glass

(including pots)

Greenhouse grapes can be grouped as follows:

SWEETWATER (S) These are early grapes, quick to mature and ripen. They are ideally suited for an unheated greenhouse. They are very sweet and juicy but without the muscat flavour. They have very thin skins when at their best but do not hang on the vine in good condition for long.

MUSCATS (M) are the finest grapes for flavour, but they need heat. They are the second group to mature and they hang well in good condition if a little warmth is provided. The fruit is firm and luscious and if allowed to hang until the skin starts to wrinkle the berries often take on the flavour of raisins. They are unfortunately difficult to get to set well unless they have warmth and some hand pollination.

VINOUS (V) These are the really late grapes and in the past were grown to provide fruit well into the new year. They are usually strong growing and crop well but need to hang for a long time in a warm greenhouse to develop their best flavour. They are not a worthwhile proposition unless heat can be provided for them in the early winter months.

The figures given are a guide to a reasonable number of bunches for a healthy vine, per 12 feet (3.6 m) of rod. A medium-sized bunch weighs about 1 lb (450 g).

Alicante (V) A black (late) grape. Vigorous grower. Fruit sets freely and needs early and severe thinning. It is inclined to form an unshapely bunch but when well thinned it can produce a handsome and imposing bunch for show with its large, almost round jet black berries covered in heavy blue/grey bloom. 9 bunches. (See p. 16).

Black Hamburgh (S) This has the best flavour of this group and is justifiably the grape best known and most widely grown under glass in this country. Excellent for pots. Of good constitution, it sets fruit freely and can be ripened well in an unheated greenhouse. If allowed to hang too

---

Opposite: 'Muscat Hamburg' (see p. 27)

long after ripening the skin becomes very thin and easily broken and the fruit deteriorates. 12 bunches.

Buckland Sweetwater (S) A round white grape carried in short broad bunches. It sets freely and ripens early particularly as a pot-grown vine. It has a pleasant flavour when in good condition but the skin thickens and the fruit deteriorates if it is allowed to hang too long on the vine. It is not a vigorous cultivar and needs to be fed well for good results. 12 bunches.

Canon Hall Muscat (M) An almost round white grape which becomes pale amber on ripening. Of excellent flavour when well grown, but it is difficult unless heat is provided at flowering time and for ripening the fruit. Berries very large. 10 bunches.

Frontignan (M) Both the black and the white Frontignan grapes are early maturing with small grapes of good quality. Well suited to growing in pots under glass. 14 bunches.

Foster's Seedling (S) The best of the white sweetwater grapes. Useful in pots. It sets freely, the bunches are of medium size and shapely, the grapes oval, juicy and of a pleasant flavour, ripening early. They should be eaten fairly soon after ripening or they will lose flavour. It has a good constitution and grows well in an unheated greenhouse. 12 bunches.

Golden Chasselas (S) A small, round, white grape giving many small bunches of sweet fruits, ripening early. Good for pots.

Gros Colmar (V) A strong growing, round, black grape, with very large fruits and handsome bunches. The skin is thick and the flavour poor until the fruit has had a long period of ripening in a warm greenhouse through the early winter. 10 bunches.

Lady Downe's Seedling (V) One of the better flavoured vinous grapes with round black berries carried in long tapering bunches. It is of good constitution but requires warmth at flowering time to ensure a good set of berries and in the early winter to ripen its fruit. 10 bunches.

Lady Hutt (S) A round white grape ripening several weeks later than others in that group. It is a robust grower, sets freely and produces large bunches of thin-skinned juicy berries. 5 bunches.

Madresfield Court (M) An early ripening oval, black grape of high quality. Its berries are large and covered with dense blue/grey bloom, the flesh firm but juicy and of good flavour, but the skin is tough. The berries are liable to split on ripening and for this reason some growers used to leave top shoots to run on without stopping at ripening to reduce sap pressure. 10 bunches.

Mrs Pearson (M) A round white grape which ripens late. It grows strongly, sets freely, bears fruit of excellent flavour but with thick skins and will hang in good condition after ripening. It does however need warmth throughout the ripening and holding period. 10 bunches. (See p. 4).

Mrs Pince (M) A strong growing, oval, black grape which produces late grapes of excellent quality but requires warmth and hand pollination at flowering to obtain a good set of fruits. It also requires warmth throughout the early winter when it is ripening. 10 bunches.

Muscat of Alexandria (M) When well grown there is no better flavoured grape. It has a good constitution but needs warmth and hand pollination if it is to set a good crop of oval white grapes. Warmth is also needed to ripen the fruit although if heat is used in the greenhouse in spring to start the vine into growth early the grapes should ripen in the September sunshine. 10 bunches.

Muscat Champion (M) A round, large, red grape, which does not keep in good quality long after ripening, but otherwise first class.

Muscat Hamburgh (M) An oval, black grape of excellent quality which ripens before most of the other muscats. It is sometimes recommended for growing in an unheated house but it is difficult to grow well. It needs warmth and hand pollination at flowering time (cross pollination, using pollen from a different cultivar is helpful). Although it is fairly vigorous in growth it has a strong tendency to shanking (see p. 64). 9 bunches.

Syrian (V) A late, oval, white grape of great vigour. It will rapidly get out of control unless regularly pruned in summer. It carries impressively large bunches but the flavour is poor unless they are allowed to hang in a warm house until Christmas. 5 bunches.

Trebbiano (V) An oval, white late grape similar in many respects to 'Syrian'. It is notable for the size of the bunches, one of which is on record as weighing 26¼ lb (12 kg). 5 bunches.

Propagation

Vine cuttings are easy to root. Soft, green cuttings taken in summer need some heat for rooting but hardwood cuttings taken in winter can if necessary be rooted outside.

Material for hardwood cuttings can be made from the prunings in December. If to be rooted in the open, cut the prunings to about 8 inches (20 cm) long, with three buds. Make a cut above the top bud and another just below the lower bud before inserting the cutting to a depth of about 6 inches (15 cm) in sandy well-drained soil. Leave cuttings rooted outside undisturbed until following autumn.

Rooting is also possible in containers in a cold frame; a cutting of two buds will be much more convenient to insert in a pot. If cuttings of this type are inserted singly in $3\frac{1}{2}$ inch (9 cm) pots in a compost consisting of two parts sand, one part loam and one part peat in January they should be well rooted by early June; then potted on into 5 inch (13 cm) pots in John Innes no. 2 potting compost.

With more sophisticated apparatus such as a heated propagating case, a single bud cutting is inserted in January in a $2\frac{1}{2}$ inch (7 cm) pot in a compost of equal parts of peat and sand and placed in the case with bottom heat of 18°C (64°F). These cuttings are made with a cut just above the bud and another about 2 inches (5 cm) below. They are pressed vertically into the surface of the rooting mixture with the bud just clear of the surface. Other methods using a horizontally inserted single bud cutting are no more effective.

Make green softwood cuttings in July or August using shoots of the current year. Cut lengths of stem about 4 inches (10 cm) long just below a bud, to include either the growing tip or at least two buds. Insert in a mixture of equal parts of peat and sand in a $2\frac{1}{2}$ inch (7 cm) pot; in a propagating case with a bottom heat of 18°C (64°F) they root readily in six or eight weeks. First move into $3\frac{1}{2}$ inch (9 cm) pots and John Innes no. 1 potting compost; when they are sufficiently robust, pot on again into 5 inch (13 cm) pots and John Innes no. 2 potting compost. In the first year the young vines, although hardy, will make better growth in a cool greenhouse.

Vines can be grafted. Some commercial UK growers of wine grapes graft their vines onto rootstocks resistant to *Phylloxera vastatrix* (see p. 57).

'Seyval', grown as a standard (see p. 52)

Vines in the Open

The grape, *Vitis vinifera*, is thought to have originated in Asia Minor and the Caucasus region. It is a sun-loving deciduous plant which does best in temperate regions where frost-free springs and warm, dry summers are experienced, with winters cold enough to induce dormancy, but not so severe as to damage or kill the plant.

The United Kingdom does not often enjoy such weather, but it is possible to grow the grape outside with reasonable success, at least in the southern half and the Midlands, given the right conditions and provided the correct cultivars are planted. In general, outdoor grapes can be planted south of a line from the Wash in the east to south Wales in the west. They will not do well in areas of high summer rainfall or strong winds or where the climate is cool throughout the growing season. In the extreme west, therefore, and the north, it is best to grow the vine on a warm wall or solid fence or to provide some kind of protection. Even in the south, the use of glass or plastic to protect against frost and as a means of increasing the temperature is very beneficial. In the West Country, for example, the growing of wine grapes in 'walk in' polythene tunnels has proved extremely worthwhile.

In essence, the United Kingdom is a marginal area for viticulture, and whilst there will be good years for the wine-maker, there will be others when there are hardly any grapes at all or the wine at best might be regarded as 'vin ordinaire'. Remember that it takes two good consecutive summers to produce a full crop. The first is needed to ripen the wood and promote fruit bud development, and the second to obtain adequate fruit set and subsequently well-ripened, fully developed bunches.

GROWING ON WALLS, WOODEN FENCES, PERGOLAS AND ARCHES

The warmer and sunnier the position, the better the quality of the fruit in terms of flavour and sugar content and undoubtedly one of the best ways of growing grapes outside in this country is against a stone or brick wall with a southerly or westerly aspect. The extra warmth and shelter the structure provides makes it possible to

---

Opposite: 'Müller Thurgau', a popular cultivar, which gives an excellent wine (see p. 50)

grow grapes in the cooler parts, where otherwise they could not be contemplated, and in the south certain outdoor cultivars which need more heat than the others can also be grown. However, it must be noted that the soil at the base of a wall can become very dry, because of the absorbent nature of the brick or stone and because the wall can cast a rain shadow over the ground in its vicinity. This means that care must be taken in soil preparation (see p. 33) and after planting to ensure that the vine is never under stress through lack of water. In addition, a wall-trained vine is more at risk from mildew and red spider mite attack owing to the hot, dry conditions.

Training methods

The vine is a flexible climbing plant and there are many ways in which it can be grown – for example, as a cordon in single or multiple form, as an espalier, as a combination of both, or even as a fan. Whichever method is used, the main point to bear in mind in training the plant is that enough space must be left between one framework branch (rod) and the next for the young laterals carrying the fruit in the summer. Vertical rods should be spaced 3½ to 4 feet (1–1.2 m) and horizontal ones 1½ feet (45 cm) apart. Basically they are pruned on the rod and spur system in the same way as the grape grown under glass (see p. 9), except that the work carried out in the

The vine in the planting hole, with the roots well spread out

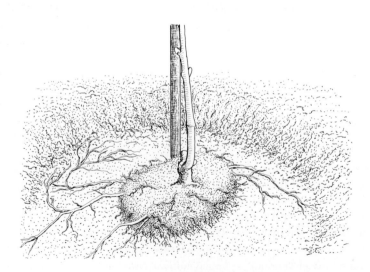

growing season comes later. A low wooden fence around a boundary, for instance, can be covered in similar fashion or the vine can be grown on the Guyot system (see p.40). A pergola or arch is best covered by growing the grape as a single or multiple cordon.

Support system
A wall-trained grape will require a supporting system of horizontal wires starting 1½ feet (45 cm) from the ground and spaced three brick courses apart, about 10 inches (25 cm). These are secured to the wall with 4 inch (10 cm) lead vine eyes driven into the vertical joints between the bricks every 3 feet (90 cm) or so. Use gauge 16 (1.6 mm) galvanised or strong plastic-covered wire. A wooden fence can be wired in a similar way. A pergola will need a system of horizontal wires at the top, spaced 10 to 12 inches (25–30 cm) apart.

Preparation and planting
Plant in the dormant season, ideally in November whilst the soil is still warm or in March when the worst of the cold weather is over. A container-grown vine can also be planted in the growing season.

Prepare the ground thoroughly. As well as being dry, the soil at the base of a wall is often poor, containing builders' rubble, for example. If necessary, replace the soil with a good quality medium loam over an area approximately 3 feet by 1½ feet by 1½ feet deep (90 × 45 × 45 cm). To improve soil moisture retentiveness, incorporate a bulky organic material such as well rotted manure or compost; about two bucketfuls should suffice. Also fork in 4 oz (110 g) balanced fertilizer like Growmore and 8 oz (225 g) bonemeal. The same preparation applies equally well to a vine against a fence or pergola.

Plant the vine to a 6 foot (1.8 m) cane 6 inches (15 cm) away from the structure and to the same depth as it was in the nursery. Spread the roots outwards away from the wall and firm the soil during the filling-in process. A container-grown plant in leaf should be watered thoroughly before it is taken out of its pot. Do not disturb the rootball except to gently tease out the perimeter roots. When planting is complete, mulch the vine generously with bark, compost or well rotted manure.

Training and pruning
If a single cordon (rod) is all that is required, the strongest shoot on the young plant should be trained up the cane throughout the summer and any other shoots kept short by pinching them back to five leaves as and when necessary. Side shoots within 12 inches

(30 cm) of the ground should be removed altogether in late summer or in early winter.

To create a multiple cordon, the first task is to grow two strong vertical shoots emanating from the main stem at about 12 to 15 inches (30–38 cm) from the ground. These are laid horizontally and tied to the lowest wire at the end of the summer and from them in future years will arise the rods of the multiple cordon (see photograph). It may be possible to establish the two shoots in the first season if the plant is strong. If weak, it is best to grow it as a single cordon in the first summer and then cut it down to 12 inches (30 cm) in November, ensuring that there remain two or three good buds at or near this point. In the second summer train vertically the two rods, keeping all other shoots pinched back to two or three leaves and then tie them horizontally at leaf-fall. If a double, rather than a multiple cordon is all that is wanted, keep the rods trained vertically but space them $3\frac{1}{2}$ to 4 feet (1–1.2 m) apart. Thereafter treat them as described below.

Subsequent pruning and training

Throughout the growing season train and tie in the rod(s) as and when necessary to fill in the available wall space. In November, after leaf-fall, any immature and unripened growth at the end of the leaders of each rod should be removed by cutting back to a bud on mature nut-brown wood. This is repeated each autumn until the required length of rod has been reached, side shoots (previous summers' growth) are cut back to one good bud near the base. This is done to form spurs along the length of the rod(s).

Pruning the established vine

Pruning, deshooting, pinching and tying in must be done regularly. Basically the method is the same as for vines under glass (see p. 9) except that the timing of the summer's operations is later.

(a) **Summer** The first task is to thin out the young laterals growing from the spurs, but not until the flowers are produced. The shoots are then thinned to one per spur or one about every 10 inches (25 cm). Retain those which are carrying the strongest flowers. The unwanted laterals are cut back to one leaf, rather than being removed completely, to reduce the risk of blind spurs. The retained laterals are stopped by pinching out the growing point at two leaves beyond the flower truss. Any sub-laterals subsequently produced are stopped at one leaf. Any barren laterals are thinned in the same way and stopped at six leaves, and their sub-laterals to one leaf.

A row of cordon vines in winter

(b) Winter As soon as the leaves are off and before the end of December, the spurs are pruned by cutting back the previous summer's laterals to one or two buds ensuring the cut is made to a plump bud. Such a bud is more likely to produce a shoot with flowers in the next summer. Thin the spurs if necessary to 9 to 10 inches (22–25 cm) apart.

The opportunity should also be taken to fill in empty wall space

35

with extension growth or suitably placed laterals. An old rod carrying too many dead spurs can be replaced by leaving in a strong replacement shoot at or near the base. In the autumn the old rod can be cut out, back to the replacement. Tie in new growth where necessary and check any old ties to ensure there is no constriction taking place. Remove any thin unripe extension growth by cutting back to strong nut brown wood. The wirework should also be examined, renewed and extended if necessary.

Watering
A vine growing against a fence or wall outside needs very little watering once it is well established, except in periods of drought. Whilst it is young, however, watering is advisable in the summer in dry periods to promote strong growth which will be better able to withstand the ensuing winter frosts. A vine growing against a house wall where the soil is usually sheltered by the eaves of the house can become very dry at the roots. Grapes grown in hot, dry conditions are more susceptible to mildew. It is essential to keep the soil reasonably moist throughout the summer by irrigation, when necessary on such sites. This involves applying about an inch of water ($4\frac{1}{2}$ gal. over 1 sq yd) every seven to ten days, in hot dry weather. A mulch of well rotted stable manure or compost applied in late February will help to conserve soil moisture.

Feeding
In late February rake away the old mulch and apply Growmore fertilizer (7%N: 7%P_2O_5: 7%K_2O) or its equivalent at 2 oz per sq yd (66 g/m²) plus sulphate of potash at $\frac{1}{2}$ oz per sq yd (15 g/m²) over the rooting area and then apply a new mulch of well rotted stable manure or compost to a depth of about 2 or 3 inches (5–7 cm).

If magnesium deficiency symptoms appear apply magnesium sulphate on a regular basis (see p. 45). With dessert grapes give the vine a liquid feed high in potassium, about $\frac{1}{2}$ gallon per plant every 10 days throughout the growing season, but stop once the berries start to ripen.

Cropping and thinning: bunches and berries
The number of bunches to allow depends upon the age and condition of the vine. As a guide, a well grown mature vine should be able to carry one bunch every 10 to 12 inches (25–30 cm) along the length of the rods. Certain cultivars which produce very small bunches with tightly clustered berries, 'Brant' and 'Cascade' for example, can be allowed more, say, at every 8 inches (20 cm).

Above: a vine trained along wires as a multiple cordon
Below: a five-rod multiple cordon in spring growth

Overcropping must be avoided. A young vine in particular will frequently remain barren for a number of years after it has been allowed to carry too many bunches. Only two or three should be left on a 3-year-old vine, perhaps four or five in the following year, and so on, provided growth is sufficiently vigorous to sustain the crop. Overcropping and the removal of too many leaves in the summer will also depress colour, sugar content and flavour.

The thinning of the berries within the bunch (see p. 15) is not necessary for wine, but could be done to advantage for dessert, provided they are cultivars which can produce large sized berries, for example 'Siegerrebe', 'New York Muscat', 'Precoce de Malingre' and the Chasselas clones. Do not thin 'Brant' and similar cultivars, as the berries are naturally small.

GROWING GRAPES IN A VINEYARD

Choosing the site

With vines grown in the open the choice of site is most important. It must be sheltered and yet in full sun and at an elevation of not more than 300 feet (90 m). The ideal site is a south or south westerly slope but level ground, or even a site sloping slightly towards the east or north is satisfactory. Avoid planting in a frost pocket as frost in the late spring can damage the blossom and young growth, possibly wiping out the potential crop for that summer. In situations where there is no alternative, the gardener should be prepared to protect the vines in some way, for example by covering them with hessian, when frost is expected.

Soils, preparation and planting

Vines are tolerant of a wide range of soils provided they are at least 12 inches (30 cm) deep and well drained. Light to medium loams are ideal but very fertile land might cause problems with excessive vigour, and extremely shallow soils over chalk with lime-induced chlorosis and poor growth in very dry seasons. Very acid soils require to be limed to bring the pH up to between 6.5 and 7.0.

Grapes root deeply and will not tolerate badly drained land nor a hard impenetrable pan beneath the surface. Heavy clay, where there are drainage difficulties, should be avoided but if this is not possible then some kind of drainage system must be installed as the first stage of soil preparation. The type depends upon the scale of planting. In a small area, where only one or two vines are planted, a simple soak-away about 2 feet square by $2\frac{1}{2}$ feet deep (60 cm × 75 cm) constructed of broken bricks, clinker and rubble would suffice, but for larger areas, say for one or more rows of vines, a length of tile or plastic drains or even a herringbone system of drains may be necessary.

The planting site should be prepared well beforehand by double digging to break up any hard layers. Ensure the area is free of weeds, and add lime if necessary. Too rich a soil is undesirable and no bulky organic manure is required unless the ground is poor.

Even then only a light dressing of well rotted manure or compost should be incorporated. In the final preparations, and just before planting, rake in a balanced fertilizer such as Growmore at the rate of 3 oz per square yard (100 g/m^2).

Plant dormant vines either in the autumn or in the spring and not in the depth of winter to avoid any risk of damage by cold. In any event, weak vines with stems of less than pencil thickness should be over-wintered in a cold frame or glasshouse and planted out when the danger of frost is over. Most plants supplied by a nursery will be on their own roots and are perfectly acceptable although a number of plants are being imported from the continent which are grafted onto rootstocks and these are usually available in the spring. Grafted vines are resistant to the vine louse, *Phylloxera vastatrix*, though fortunately it is very rarely a problem in this country. Some rootstocks are more suited to certain soil types than others (see table).

Suitability for soil types					
Shallow poor, stony, dry	Deep fertile without chalk	Deep fertile with chalk	Heavy clay with chalk	Heavy clay	Chalk
5BB 125AA	5C SO4 125AA	5C SO4 5BB 125AA	SO4	125AA 5BB SO4	SO4

Firm planting is essential and the roots of the young vines (often pot grown) must be carefully spread out to encourage quick establishment. Subsequently mulch them with a little rotted manure or compost so that the lower buds are covered for the winter. With grafted plants the union should certainly be covered in winter. In the spring the mulch should be gently pulled away from the stem. The graft union, indicated by a slight bulge, must be about 1 inch (2.5 cm) above soil level to prevent scion rooting because if this occurs the desired effect of the rootstock on the vine will be lost and the plant will no longer be resistant to *Phylloxera*.

Training methods

There are many methods by which a vine can be trained in the open. It could for example be grown in a sunny corner as a vertical cordon, trained up a stout pole, or as a line of cordons supported by canes and wires, or even as a clean stemmed standard like a weeping rose (see photo on page 28). It could also be grown in a container (see pp. 21–23). The grape in any of these examples would be spur pruned in the winter and the laterals pinched back to three or four leaves beyond the bunch in the summer to keep the plant in shape.

For vineyards large and small, however, the most widely used method of training is the Guyot system in single or double form. Basically this is a replacement system of pruning, whereby the wood which has fruited in the previous summer is cut out each winter and new wood tied in to fruit in the next summer. The Single Guyot has one fruiting arm whereas the more popular Double Guyot has two. The latter is preferred by most growers as it takes fewer plants for the same amount of space and production is about the same. There is a case for the Single Guyot where there is no room for two arms, where the soil is very poor or where a large number of cultivars are wanted relative to the land available. In either form, the arms are trained fairly close to the ground to take full advantage of the reflected and radiated warmth from the soil.

Support

Vines grown in rows in the open require a stout fence of posts and wires. For the Guyot system, the posts should be spaced 13½ to 15 feet (4–4.5 m) apart depending upon the spacing of the plants within the row.

The end posts should be 7 feet (2 m) long, by 3 inches (7.5 cm) diameter, and 7 feet (2 m) by 2 inches (5 cm) diameter for the intermediate posts, driven 1½ feet (45 cm) into the ground.

The wood must be preserved against rot, preferably impregnated under pressure with a preservative or dipped. The fence should be strutted at each end. Galvanised fencing wire is used for the wirework. The two lowest wires are single and the upper wires double. Secure the wires to the end posts with straining bolts and to the intermediates with staples. Do not drive the staples completely home, but allow the wires to run freely through them. The two lower wires (12 gauge/2.5 mm) are set at 15 inches and 21 inches (40 cm and 55 cm) from the ground and the upper wires (14 gauge/2 mm) at 3, 4 and 5 feet (90 cm, 1.2 and 1.5 m) (see p. 47). Preferably the rows should run north to south to reduce mutual shading. If one side of each of the three sets of double wires can be taken down at the time of winter pruning, the task is made easier. To do this, fix a short length of chain to the ends of the wires and instead of wire staples on the intermediates, use small cup hooks.

The chain is pulled tight around the end post so that the projecting end of the straining bolt is caught in one of its links. The wire itself is not attached to the bolt (see below).

Spacing *(Double Guyot)*
Space the vines 4½ to 5 feet (1.3–1.5 m) apart in the row. Allow the wider spacing on good soils. The spacing between the rows is 5 to 6 feet (1.5–1.8 m). If a grass sward is preferred, then space the rows 6 feet (1.8 m) apart with a clean grass-free band of 18 inches (45 cm) wide maintained down the row.

Planting
The young vines should be planted to a stout cane 6 feet (1.8 m) in height. Plant to the same depth as the vine was in the nursery, being careful not to bury the graft union under the soil's surface where grafted plants are used. See Weed Control p. 46, for details of planting through polythene.

Initial training
Normally, the vine will break into growth some time in May. In the first summer after planting allow the strongest shoot, usually the topmost, to grow and train it vertically up the cane, all others being cut out. In November prune this shoot to a bud just below the bottom wire ensuring that there remain at least two more good buds beneath it. In the second summer train in three shoots, tied vertically, pinching back any others to one leaf.

A simple way of securing wire at the end using a chain link over a straining bolt

At the end of the second summer after leaf fall, start training to the Double Guyot system. One shoot is tied down to the left and one to the right onto the bottom wires. The third shoot is cut to three buds and this will provide the three replacement shoots for the next year (see photographs). The arms should not overlap with their neighbours, but should be cut to a bud short of the arms on the neighbouring vine.

A weak vine less than 2 feet (60 cm) high at the end of the first year should be cut down to three buds and grown once again as a single stemmed plant in the next growing season and then pruned as above.

Pruning the established vine (see opposite)

November. Prune as soon as possible after leaf fall, but no later than the end of January. However, early winter pruning stimulates earlier bud burst (by about seven days) than vines pruned in the spring. Possibly where vines are planted in areas prone to spring frosts, there could be an advantage in pruning in March or April in the hope that they may escape frost damage, but it must be noted that late pruned vines can 'bleed' badly from the pruning cuts. Nevertheless, provided the vines are healthy, the bleeding does not appear to weaken the plants and the flow stops once the plant has developed some leaves. Note, there are, or should be, three strong replacement shoots trained up the cane or post. If not, utilise three strong laterals, nearest to the centre. The two arms which bore fruit last summer are cut back to the replacement shoots. One shoot is then tied down to the left and another to the right arching over the lower wires and the third is cut to three or four buds. These buds will provide the replacement shoots for next year.

January. Winter pruning completed. Replacement shoots arched over and tied to the lower wires and immature wood cut off, leaving 2 to 2½ feet (60–75 cm) each side. Third shoot cut to three or four buds.

July, August, September. The fruit carrying laterals are trained through the double wires. Prune them to two or three leaves above the top wires as, and when, necessary, using secateurs or shears. Remove any sub-laterals. The replacement shoots (three for a Double Guyot, two for a Single Guyot) are trained up the cane or post and stopped at about 5 feet (1.5 m). Sub-laterals are pinched to one leaf, and as a counsel of perfection, any blossom on the replacement shoots should be removed, though this is not strictly necessary if the growth is strong. Remove any surplus shoots or suckers coming off the main trunk below the two arms.

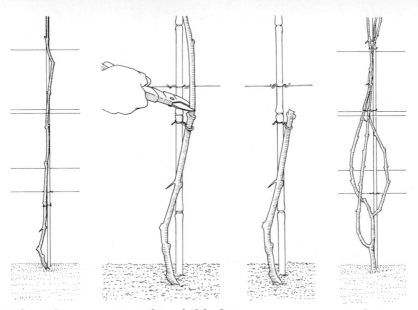

Left to right: a young vine at the end of the first summer, grown as a single stem; the young vine in autumn is cut to a bud just below the bottom wire; the young vine, with pruning completed; and the same vine in the autumn of the second year, when three upright shoots have been allowed to grow.

Starting in early September, gradually remove a *small* proportion of the lower leaves, just sufficient to expose the bunches to more sunlight and improve air circulation. Be careful not to expose the bunches too suddenly to the sun as this may lead to sun scorch. This reduction of foliage will help in the ripening of the berries and the control of botrytis.

Cropping
In the first year of fruiting, two or three bunches only should be allowed on a three-year-old vine, four or five in the next year, and full cropping thereafter, provided that growth is healthy and vigorous enough to sustain the crop.

Feeding and watering
In February apply Growmore fertilizer or equivalent at 2 oz per sq yd (66 g/m^2) plus sulphate of potash at $\frac{1}{2}$ oz per sq yd (15 g/m^2) along each side of the row to a width of 1 foot (30 cm) on either side. On light soils in the spring mulch the vines with a dressing of mushroom compost, bark, well rotted manure or garden compost to a depth of about 2 inches (5 cm). Vines are prone to magnesium deficiency, sometimes shown by yellowing of the leaves in

43

summer, but not to be confused with chlorosis due to alkaline conditions. If magnesium deficiency symptoms occur, apply a foliar spray of ½ lb magnesium sulphate (coarse Epsom salts) in 2½ gallons of water (220 g/11 litres) plus a few drops of mild liquid detergent and repeat the spray 14 days later. Thereafter apply the magnesium sulphate as a top dressing in the late spring at 2 oz per sq yd (66 g/m²) (see also p. 36).

Grapes grown for dessert will benefit from liquid feeds high in potassium in the growing season. Apply the fertilizer once a fortnight from the time the berries start to form until they begin to ripen.

Well established vines are usually able to cope with dry weather except on very shallow soils, nevertheless it is advisable to water them in drought conditions.

Spray programme
The two most important diseases on grapes grown out of doors are powdery mildew (p. 60) and botrytis (p. 59) and a regular spray programme to control them is advised. Details of pests and diseases which may attack outdoor vines are given on pp. 55–61.

The two groups of sprays that should be applied according to the manufacturer's instructions as a routine are as follows:

In mid-June: benomyl, carbendazim or thiophanate-methyl, or wettable sulphur or sulphur dust for powdery mildew control. Repeat four times at fortnightly intervals. (N.B. the minimum interval between spraying and harvesting is seven days).

Usually just before the flowers open, but adhere to the manufacturer's instructions in this regard: benomyl, carbendazim or thiopanate-methyl for botrytis control. Repeat every 14 days until three weeks before harvest.

Weed control
It is important to control weeds in the first year of planting when the vine is trying to become established, as well as in subsequent

---

Top left: an established vine trained on the Double Guyot system before pruning
Top right: remove arms back to young replacement shoots
Centre left: after the arms are removed, a number of replacement shoots remain to be dealt with
Centre right: three replacement shoots are left. The central one is cut back to three buds, each of which will produce a replacement shoot next summer
Below left: pruning is complete, all that remains to be done is to tie the arms to the lowest wires
Below right: a Double Guyot in spring growth

years when the plant is bearing a crop. Weeds compete for water and nutrients and also are liable to create humid conditions around the bunches, making fruits more prone to grey mould. As the methods of weed control in vines are limited to hoeing or careful use of the contact-action weedkiller mixture paraquat with diquat, it is essential to weed-kill regularly at intervals throughout the spring and summer months. This is necessary to prevent quick-developing annual weeds maturing and setting seeds and to ensure that no perennial weeds have time to become established.

Paraquat with diquat will kill any green tissue, green bark, green buds or leaves on contact. It has a burning scorching action and should be directed onto the weeds, but must be kept clear of vines. It will kill annual weeds and check perennials, but the latter will usually survive to resume growth.

In no circumstances attempt to use the growth-regulator (hormone) type of weedkillers near vines as accidental contact of spray or spray drift can cause severe distortion.

Black polythene laid over the ground as a weed suppressant and as a mulch to conserve moisture is widely used by commercial growers and is worth adopting by the amateur. One can either use solid 500 gauge (125 mμ), 3 feet (1 m) wide polythene or perforated or woven plastic. The latter two allow the rain to percolate down to the roots, but not the weeds to grow through. Lay the polythene

Summer pruning a Double Guyot by cutting back lateral shoots to two or three leaves above the topmost wire

along the row, cut slits or burn holes at each planting station and plant the vine through the aperture. The ground should be thoroughly prepared and in good condition as for normal planting and be well watered, not dry. Bury the sides of the polythene vertically about 6 inches (15 cm) deep to prevent it blowing away. It is, of course, also possible to use plastic as a weed suppressant and mulch around vines already planted, but here two 20 inch (50 cm) wide sheets are necessary, one on either side of the plants allowing about a 4 inch (10 cm) overlap along the middle.

Diagram of the pruning of a mature vine

Winter: Before pruning

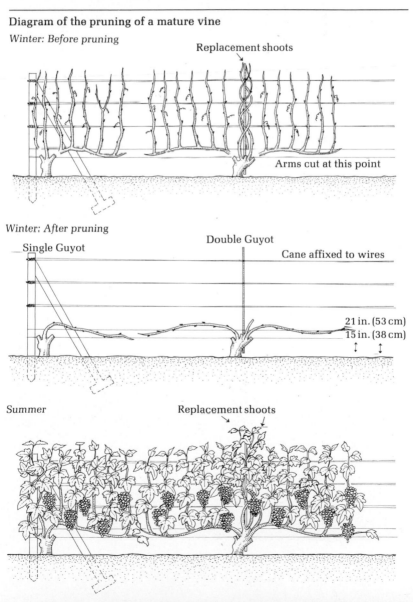

Replacement shoots

Arms cut at this point

Winter: After pruning

Single Guyot

Double Guyot

Cane affixed to wires

21 in. (53 cm)
15 in. (38 cm)

Summer

Replacement shoots

— Cultivars for Growing Outside —

Black grapes, other than a few hybrid cultivars, do not ripen sufficiently well when grown outside to make a wine of high quality, except perhaps in very hot summers. With one or two exceptions, they are best therefore grown on a wall or fence or given some protection for at least part of the growing season. For this reason in most commercial vineyards in Britain, white grapes predominate. In general, white cultivars which ripen in early to mid October in the south are preferred as the early ones are very liable to be attacked by wasps. Further north early cultivars should be chosen rather than the late ones which might not ripen.

Besides grapes for wine, there are a few outdoor cultivars also suitable for dessert in a good summer and these are listed below. Nevertheless, as might be expected, they will acquire a sweeter flavour if grown against a warm wall. Cultivars normally recommended for under glass are unsuitable.

The season of ripening given below refers to the south of England.

WHITE

Bacchus (mid season) Similar to 'Müller Thurgau', but more reliable in cropping. Ripens its wood well. Gives a good quality wine.

Chardonnay (late) A first quality white wine and Champagne grape which does best on light chalky loams. It is not suited to heavy clay. Yield moderate.

Chasselas (late) Of which there are a number of clones, e.g. 'Chasselas d'Or', 'Chasselas 1921', 'Chasselas Rose Royale', (see p. 18) and 'Royal Muscadine'. All yield good berries excellent for dessert and wine: however, this cultivar is too late in the open and is better on walls. Late October.

Himrod Seedless (mid-season) Good for dessert, but needs a warm situation and best on a wall. Berries small, yellow. Flavour sweet, good. Large loose bunches.

Huxelrebe (early mid-season) Vigorous. Heavy cropper and needs thinning on a young plant. Susceptible to wasps and botrytis. Produces a high quality muscat flavoured wine.

Madeleine Angevine (early) A heavy cropper. Vigorous. Prone to mildew. Relatively hardy. Suitable for colder areas. White wine. Quality only fair. Late September to early October. (See p. 1).

Opposite: 'Bacchus'

Müller Thurgau (Riesling Sylvaner) (mid-season) The most widely planted. It has a delicate flavour and gives an excellent wine. Passable as dessert. Needs good weather during pollination. Mid October. Highly recommended. (See p. 30).

New York Muscat A tawny coloured grape of excellent muscat flavour in a hot summer. Must be grown on a wall. Large berries. Mid to end October. Dessert.

Perle de Czaba (early) An excellent muscat flavoured berry. Best on a wall or under glass. Weak to moderate vigour. Light cropper. Late September. Dessert.

Precoce de Malingre (early) A large berried sweetwater grape of good flavour. Moderate vigour. Light cropper. Dessert and wine. late September.

Reichensteiner (early mid-season) Moderately vigorous. Crops well. Grapes high in sugar, but makes a neutral wine and best blended. Some resistance to botrytis.

Scheurebe (mid to late) Moderately vigorous. Moderate yield. Makes a good quality wine with a flowery aroma.

Opposite: 'Seyval', a good hybrid which is easily grown outside in Britain
Below: 'Müller Thurgau' grown on a wall

Schonburger (late) Moderately vigorous. Makes a good quality wine, but requires a warm sheltered site to crop well.

Seyval (Seyve Villard 5.276) (mid-season) A hybrid, some resistance to powdery and downy mildew. A heavy cropper and easy to grow. Blends well with 'Müller Thurgau'. Recommended. Mid to end October. (See p.28).

Siegerrebe (very early) A golden brown berry of good flavour with a trace of muscat and high sugar content. Prone to wasps. Dessert and wine. Medium vigour. Late August to early September. (See p. 54).

BLACK

Brant (late) A hybrid of Canadian origin often grown for its autumn colour, bears heavy crops of small, sweet black grapes, liked by children. Some resistance to mildew. Vigorous, useful wall cover. Mid October.

Cascade (Seibel 13.053) (mid-season) Resistant to powdery mildew. Low in sugar and high in acid. A fair quality wine. Small bunches of deep purple berries with red juice. Very vigorous, excellent as a wall cover plant. Early October.

'Cascade', a vigorous black vine, ripening in mid-season

Leon Millot (mid-season) A French hybrid with good resistance to mildew. Small bunches of blue-black berries. Very vigorous. Makes a fair quality red wine. Crops well. Reliable. Early October.

Muscat Bleu A muscat flavoured berry of good size. Bunches medium. Excellent flavour but seeds are large. Best on a wall, but will ripen out in the open in a warm summer. Attractive autumn colour. Mid October. Dessert.

Noir Hatif de Marseilles (early) Small to medium sized berries and bunches of slight muscat flavour. Good quality. Rather a weak grower and requires fertile soil. Needs protection or a warm wall. Late September. Dessert.

Pinot Noir (late) A very high quality wine grape used for champagne. In most years in Britain it needs the warmth of a wall, glass or a polythene tunnel to ripen the berries fully. The wood of the vine itself matures well and flower initiation is good, even in a cool summer.

Pirovano 14 Medium sized bunches of large berries of good non-muscat flavour suitable for dessert and wine. Best on a wall or with protection. Mid October.

Schuyler (mid-season) Fruits black, large berries and bunches. Of American origin, but is a European type without any slipskin character. Flavour sweet, slightly muscat. Dessert and makes a fair wine. Vigorous. Crops well outside and against a wall.

Tereshkova (early) A hybrid with medium sized berries of purple-red with delicate muscat flavour. Attractive autumn colour. Late September. Dessert. (See p. 54).

Triomphe d'Alsace (early) Very vigorous, crops heavily. A hybrid which is resistant to mildew. Makes an intensely coloured red wine of fair quality. Suitable outside and against a wall.

PROMISING CULTIVAR

Roem van Boskoop (Gloire de Boskoop or **Boskoop Glory)** A Dutch sweetwater black grape of good dessert quality for an outdoor grape. Bunches and berries of medium size. Crops well. Moderate vigour. Ripens late September to early October.

AMERICAN VINES

These are North American hybrids often with the vine species *Vitis labrusca* in the parentage. There are a number of cultivars e.g. 'Concord' (black) and 'Fragola' (pink). Small or large bunches of good sized berries, having a thick slippery skin and a taste faintly reminiscent of strawberries. Sometimes described as having a foxy flavour which is imparted into wine made from them. Best grown against a wall. Good resistance to mildew. Late October.

Above: 'Siegerrebe', a very early grape, suitable for dessert and wine
Below: 'Tereshkova', a hybrid which produces attractive autumn colour

Grapes grown out of doors are generally much less liable to be attacked by pests troublesome in greenhouses such as mealybugs, scale insects, red spider mite and whitefly. However, an outdoor vine growing in a warm sheltered position, such as against a south facing wall, may also have these problems. When applying pesticides there is a danger that the bloom on glasshouse grapes will be marred if sprays are applied after the fruits have begun to swell. As far as possible, pests should be controlled before this stage is reached, or formulations less likely to affect the bloom, such as smokes or aerosols, should be used. Insecticides should not be used within 10 days of picking the fruit.

Some of the pests listed below have a wide range of host plants. If the vine is being grown in a greenhouse with other plants, these should also be checked for the presence of pests and treated accordingly.

Birds. Birds are the most damaging pests of outdoor grapes and they may eat the entire crop unless the vine is covered with birdproof netting. This will need to be in place from the beginning of September to the end of October depending on the cultivars grown. Netting with a mesh of $\frac{3}{4}$ inch or an inch (19–25 mm) must be used. If the vine is too large to be netted completely, individual bunches can be enclosed in bags made from muslin or nylon stockings.

Glasshouse red spider mite. Red spider mites are tiny, eight-legged, sap-feeding animals that occur in large numbers on the underside of the leaves. They overwinter as adult mites and they begin feeding on the new foliage during April or May.

The first sign of an attack is a fine yellowish green speckling that can be seen on the upper leaf surface; examination of the underside of the leaf with a hand lens should reveal the mites. As the infestation increases, leaves begin to dry up and drop off, and the vine may become covered in a fine silken webbing which is produced by the mites. Despite their common name, the mites are not really red in the summer and they range in colour from nearly black to yellowish green or orange.

Early treatment is required to prevent a damaging infestation from developing. As soon as mite damage is seen, spray the vine with malathion, repeating at intervals as necessary. Pesticide-resistant strains of the mite may occur so as an alternative red spider mite can be controlled by introducing a predatory mite called *Phytoseiulus persimilis*, although this cannot be used if insecticides are being used to control other pests. Supplies can be obtained from: Natural Pest Control (Amateur), Watermead, Yapton Road, Barnham, Bognor Regis, West Sussex PO22 0BQ; English Woodlands Ltd, Hoyle Depot, Graffham, Petworth, Sussex GU28 0LR; Biological Crop Protection, Occupation Road, Wye, Ashford, Kent TN25 5AH.

Glasshouse whitefly. Adult whiteflies are tiny moth-like insects with white wings which live on the underside of the leaves. The nymphal stages are flat, whitish green scales, and both the adults and nymphs feed by sucking sap. Like mealybugs, whitefly can create problems with honeydew and sooty mould. The eggs and nymphal stages are relatively immune to insecticides, so frequent sprays are necessary to kill the adults as they develop.

Sprays or smokes containing permethrin or a related compound, such as pyrethrum, are effective provided resistance has not occurred. As an alternative to insecticides, a parasitic wasp called *Encarsia formosa* can be introduced during the summer. This is available from the same sources as *Phytoseiulus* (see above).

Mealybug. This is a common pest of glasshouse grapes. Mealybugs are soft bodied, pinkish grey insects up to $\frac{1}{8}$ inch (3 mm) long, and they suck sap from the leaves and stems. They tend to cluster in the leaf axils and they secrete white woolly fibres around themselves. They excrete a sugary substance called honeydew that makes the vine sticky and encourages the growth of a black sooty mould on the surface of the foliage and fruits.

The winter tar oil treatment described below under scale insects also deals with mealybugs. If they persist during the growing season they can be controlled by thorough spraying with malathion at 14 day intervals during the spring and early summer. A ladybird beetle, *Cryptolaemus montrouzeri*, can be obtained from the addresses above.

Scale insects. Scale insects feed by sucking sap from the vine and the type most commonly found is called brown scale. The mature insects are covered by hard, shiny brown, convex shells about $\frac{1}{4}$ inch (6 mm) long and they are attached to vine stems. Apart from the recently hatched nymphs, scale insects do not move once they have found a suitable feeding place.

A less common but more spectacular type of scale insect is the woolly vine scale. The mature females of this scale have dark brown, wrinkled shells about $\frac{3}{16}$ inch (4 mm) in diameter, which are perched on the edge of a mass of white waxy threads which contains the eggs.

Scale insects can be controlled by thoroughly brushing the vine rods with a tar oil wash during December. Tar oil will damage any plants that are in leaf, so care must be taken if other plants are growing nearby. Before treatment, the vine rods should be scraped to remove loose bark and as many scales as possible. Collect and burn the scrapings. If necessary, the vine can also be sprayed with malathion in late June and mid July as the young nymphs begin to hatch.

Vine blister mite (Erinose). This is a microscopic mite that can occur on both glasshouse and outdoor grapes. It overwinters inside the buds and starts feeding on the leaves in the spring. As a result of this feeding, the upper surfaces of the leaves develop raised patches and the undersides of these blisters are covered by fine white hairs. These hairs may be confused with mildew disease but the blistering effect is diagnostic for the mites. The symptoms first appear in May or June, and as the summer progresses the hairs darken and become reddish brown. The mites live among the hairs and feed on the leaf surface but, apart from some disfigurement to the leaves, they seem to cause no real harm. Infestations can usually be dealt with by picking off the affected leaves.

Vine phylloxera. This serious pest of vines is rare in this country although outbreaks occasionally occur when infested vines are imported into Britain. It is an aphid-like insect with a complex life cycle, having different forms that cause galls on the leaves and roots. The leaf galls are the most obvious symptom. They are pinkish or yellowish green, swollen, rounded structures that form on the leaf surface. The galls are hollow and contain the pest. The root infesting forms cause swellings on the roots and heavy infestations will kill the plant unless the vine has been grafted onto a phylloxera-resistant rootstock. Suspected outbreaks must be reported immediately to the local office of the Ministry of Agriculture who will take the appropriate control measures.

Vine weevil. The adult weevils feed at night and eat irregular shaped notches from the leaf margins of many plants, including grape vine. The beetles are mainly black with small brown patches on the wing cases. The slow moving beetles are about $\frac{3}{8}$ inch (9 mm) long and they have elbowed antennae. Their larvae are fat, creamy white, legless grubs with brown heads. They live in the soil and feed on roots. Well established vines are unlikely to suffer serious damage, but pot-grown or young vines may be killed by the grubs. If damage caused by the grubs or adults is seen, they can be controlled by drenching the soil with a spray strength mixture of HCH (BHC). A nematode treatment for vine weevil grubs is available from Biological Crop Protection, address on p.56.

Wasps. In some years wasps are very numerous and they will damage large numbers of grapes, especially of the early ripening varieties. An effort should be made to find and destroy as many wasp nests as possible in the locality. It is easier to follow the wasps' flight paths as the sun is setting and likely places to look for nests are ditches, hedge bottoms, and under the eaves of buildings. Once located, the nests can be destroyed by placing a wasp powder such as carbaryl in the entrance hole. Individual bunches of grapes can be protected from wasps by enclosing them in bags made from muslin or old nylon stockings. Wasps can also be kept out of greenhouses by screening the ventilators with nylon netting.

Fruit protected from wasps and birds by an old nylon stocking

Diseases and Disorders

DISEASES

Downy mildew occurs only occasionally in this country, mainly on outdoor grapes. It shows as light green blotches on the upper surface of the leaves and as a downy mildew on the lower surface of these patches. Affected areas dry up and become brittle causing the leaves to curl and fall. Diseased berries shrivel and become brown and leathery. The tips of the shoots may also be attacked. Remove and destroy all diseased tendrils and leaves to remove the overwintering stage of the fungus, although some spores also overwinter in the bud scales and on shoots. Where infection is expected to occur, apply a protective spray of mancozeb, but not when the vines are in flower, and repeat at 10 to 14 day intervals, ceasing when further spraying would cause an unsightly spotting of the fruit. (See also vine blister mite, p. 57).

Grey mould (botrytis) is the most troublesome disease on outdoor vines, but it can also be a nuisance on indoor grapes in poorly ventilated greenhouses where humidity is high.

Affected grapes rot and become covered with a dense brownish grey furry mass of fungal growth. The fungus either attacks the berries through wounds or it may invade the floral parts so that some of the fruits are already diseased as they develop. By whichever method infection occurs, once the disease is established it can spread rapidly by contact and also by means of air-borne spores which are produced in vast numbers. In wet weather and in very humid greenhouses crop loss can be considerable. Under glass the trouble can be prevented to a certain extent by adequate ventilation to reduce the humidity and by the prompt removal of unhealthy berries and leaves. Once the disease has occurred fumigate the greenhouse with tecnazene smokes.

It is difficult to control the disease on outdoor grapes so in wet seasons try to improve the aeration around the bunches by thinning out some of the berries and judicious removal of some of the shoots or leaves. Sprays of benomyl, carbendazim or thiophanate-methyl will control grey mould for a season or two, but regular use of these fungicides may lead to the development of strains of the fungus which are resistant to them so that they will become ineffective. For those growing grapes on a commercial scale, fungicides are available to control this disease, and advice should be sought on this subject.

Honey fungus. Both indoor and outdoor vines are very susceptible to infection by this soil-borne fungus which can kill affected plants very rapidly. White fan-shaped growths of fungus develop beneath the bark of the roots and main stems at and just above ground level. Brownish black root-like structures known as rhizomorphs may be found growing on diseased roots; these grow out through the soil and spread the disease. Control of the fungus is difficult and it is essential, therefore, to trace the source of infection so that all woody debris can be dug up and burnt together with dead and dying vines and as many roots as possible. (For more detailed advice, consult the RHS Garden Wisley.)

Powdery mildew is most troublesome on indoor vines particularly in a cold greenhouse especially if the soil is dry and the atmosphere

Outdoor grapes affected by grey mould

is humid or stagnant. Powdery mildew is also common on outdoor grapes, particularly on those growing in very dry positions such as against walls. A white powdery coating of fungus spores develops only sparsely and the most obvious symptom is a grey or purplish discoloration of the diseased areas. The disease can also attack the flowers and fruits causing them to drop or at a later stage the grapes may become hard and distorted and split, and are then often affected by secondary fungi, such as grey mould (see p. 59), which cause extensive rotting.

Powdery mildew can be prevented to a certain extent by mulching and watering to stop the soil from drying out. As soon as the disease appears apply benomyl, carbendazim or thiophanate-methyl; dusting with sulphur can also be carried out. If the disease has been troublesome in previous years the first application of fungicide, whichever is used, should be given 10 to 14 days before mildew is expected. Four applications during the season may be needed to keep the disease in check. In a cold greenhouse in dull weather, it may be necessary to provide some heat temporarily so as to avoid too humid an atmosphere and careful but sufficient ventilation should be given to get good air circulation. Over-crowding of the shoots and leaves must be avoided to prevent stagnant air conditions.

DISORDERS

'Bleeding' of the wounds. Often as a result of late winter pruning, the sap flows freely from the pruning cuts, especially if the wounds are large. The sight of such 'bleeding' can be disconcerting to the inexperienced grower. There is no need for concern, however, as provided the vine is healthy, the sap loss does not appear to harm the plant and it will eventually stop in the spring when the buds break into growth. There have been many remedies advocated in the past to staunch the flow such as cauterising the wound, applying sealing wax, carpenter's knotting etc., but these rarely work once the 'bleeding' has started. The real answer is not to prune late, but to complete the job by the end of December, and, preferably, immediately after leaf fall.

Exudations. A common phenomenon in the spring is the presence of transparent globules resembling eggs, on the lower leaf surfaces and petioles. These small round greenish or colourless droplets are not the eggs of any pest but are due to a natural exudation from the plant. This type of exudation commonly occurs on young growth and indicates that the root action is very vigorous and the plant is in good health. The symptoms are inclined to be more obvious, however, on plants growing under glass where the atmosphere is humid.

Fruit splitting. This trouble most commonly occurs as a result of an attack by powdery mildew (see p. 60). However it is occasionally due to irregular growth such as may occur when heavy rain follows a period of drought. Affected grapes usually rot as a result of subsequent grey mould infection. There is very little which can be done to prevent this trouble apart from mulching to conserve moisture and by watering in dry periods before the soil dries out completely.

Magnesium deficiency. Vines are soon affected by a soil deficiency of magnesium, symptoms being produced in the leaves. These usually show as a yellowish orange discoloration between the veins, but in some cultivars the leaves may have purplish blotches. Later the affected areas turn brown; these symptoms should not be confused with sun scorch (p. 64). The deficiency can be corrected by spraying the foliage at the first signs of trouble, with $\frac{1}{2}$ lb (220 g) of magnesium sulphate in $2\frac{1}{2}$ gal (11 litres) of water plus a spreader such as soft soap or a proprietary product. Apply two or

Magnesium deficiency symptoms

62

Left: powdery mildew (see page 60)
Right: some berries may fail to develop, a condition known as shanking (see p.64)

three times at fortnightly intervals. Soil applications act more slowly, but are useful as a long-term cure. For outdoor grapes apply the magnesium sulphate in subsequent years as a soil dressing in late spring at 2oz per sq yd (66 g/m²).

Oedema is a physiological disorder which is more likely to occur on vines grown under glass. Fairly late in the season, small wartlike outgrowths appear on the stalks when the fruits are developing and sometimes on the grapes themselves and even on the lower leaf surfaces. These outgrowths may break open and then have a blister-like or white powdery appearance, or they may become rusty coloured and appear as brown scaly patches. The trouble is caused when the roots of an affected plant take up more water than the leaves can transpire and this may be due to extremely moist conditions either in the soil or in the atmosphere or both. Once oedema has occurred, do not remove the affected parts as this will only make the trouble worse. No special treatment can be given and the remedy is to maintain drier conditions both in the air and soil; with correct cultural treatment the affected plant should recover in due course.

Scald and scorch. Scald which shows as discoloured sunken patches on the berries, and scorching of the foliage which results in large pale brown patches, is due to hot sun striking through glass on to moist tissues. These troubles usually occur where the ventilation has been poor, but light shading may also help to prevent them. Remove affected berries and leaves.

Shanking first shows at the early ripening stage when odd berries or small groups of berries do not colour and develop naturally. It starts as a dark spot along the stalk of the grape which is finally girdled so that it shrivels and the grape stops developing and fails to ripen. The grapes are watery; black grapes turn red and white grapes remain translucent. The fruit if tasted will be found sour and unpalatable.

This disorder is usually a result of one or more unsuitable cultural conditions, such as faulty root action, due to too wet or too dry soil conditions, or the penetration of stagnant soil by the roots, or over-cropping of the vine which puts an undue strain upon the root system. The remedy, therefore, is to study the soil conditions to see whether the roots could have been affected by drought or waterlogging. Mulching or resoiling may be necessary if there is any sign of such damage. At the same time reduce the crop for a year or two until the vine has regained its vigour. When shanking occurs fairly early in the season it is sometimes possible to save the rest of the crop by cutting out the withered berries and spraying the foliage with a foliar feed, providing the vine is not being overcropped by taking too many bunches.

Spray damage. Damage by sprays, usually due to the misuse of weedkillers (such as 2,4-D; mecoprop), occurs on both indoor and outdoor vines. Affected leaves become narrow and fan-shaped, are frequently cupped and the shoots twist spirally. Affected plants usually grow out of the symptoms in due course, but it is better to avoid such damage by the careful use of hormone weedkillers. Keep special equipment for their use, do not spray on a windy day nor use within 400 yards (364 m) of outdoor grapes. Close all ventilators in greenhouses if any spraying is to be carried out in the vicinity. Do not leave weedkillers in a greenhouse as vapours from them may affect plants on a hot day.